Series / Number 90-011

Two-Way Communication: Political and Design Analysis of a Home Terminal

NOAM LEMELSHTRICH

Center for Policy Research

AUTHOR'S NOTE: The preparation of this paper was supported by a research grant (no. GI-29940A1) from the National Science Foundation for a study titled "Minerva: A Participatory Technology System," which is still in progress at the Center for Policy Research.

SAGE PUBLICATIONS / Beverly Hills / London

Printed in the United States of America

For information address:

SAGE PUBLICATIONS, INC.
275 South Beverly Drive
Beverly Hills, California 90212

SAGE PUBLICATIONS LTD.
St George's House / 44 Hatton Garden
London EC1N 8ER

International Standard Book Number 0-8039-0464-9

Library of Congress Catalog No. L.C. 74-79048

FIRST PRINTING

When citing a professional paper, please use the proper form. Remember to cite the correct Sage Professional Paper series title and include the paper number. One of the two following formats can be adapted (depending on the style manual used):

(1) LEMELSHTRICH, N. (1974) "Two-Way Communication: Political and Design Analysis of a Home Terminal." Sage Research Papers in the Social Sciences (Policy Research Series, No. 90-011). Beverly Hills and London: Sage Publications.

OR

(2) Lemelshtrich, Noam. 1974. *Two-Way Communication: Political and Design Analysis of a Home Terminal.* Sage Research Papers in the Social Sciences, vol. 2, series no. 90-011 (Policy Research Series). Beverly Hills and London: Sage Publications.

CONTENTS

PREFACE

Most of us are so immersed in the existing social arrangement it is difficult for us to *seriously* contemplate the social consequences of radically different systems. But societies do change fundamentally, even if not always rapidly or dramatically. Earlier generations moved from the country side and life, into urban centers. More recently, middle classes and corporations escaped the city. Lemelshtrich deals with a technology which might help allow people to use the places they live–small town, suburb, villages–as centers of knowledge, communication, and decision-making. Equipped with two-way communications and retrieval terminals, each home could be a push-button away from memories and knowledge stored in computers anywhere, and–free to dialogue with persons. In short, each home could have a branch of the Library of Congress, a Corporate Headquarters data book, and–executive conference room.

Not everyone may welcome this "de-differentiation" of home and work (or home and other areas of decision-making, e.g., political). But no one can afford not to know more about the technologies which could help bring about such a fundamental change.

Amitai Etzioni

NOAM LEMELSHTRICH was a member of the M.I.T. task force for the development of telecommunication policy from 1971 to 1973, and will begin teaching at the Hebrew University in the Communications Department in 1975. He received his B.S. from California State University at Northridge in 1970, his M.S. from Stanford in 1971, and his Ph.D. from M.I.T. in 1973. His "Screen Feedback Home Terminal for Two-Way Cable T.V." appears in Ithiel de Sola Pool (ed.) *Talking Back: Citizen Feedback and Cable Technology*, published in 1973.

Two-Way Communication: Political and Design Analysis of a Home Terminal

NOAM LEMELSHTRICH

The Hebrew University of Jerusalem

Introduction

Recent developments in communication technology will make it possible for TV viewers to indicate a response to the information they receive through this medium. The possible sociopolitical implications of such a two-way interaction are widely discussed in recent literature on Cable Television (de Sola Pool, 1968; T. B. Sheridan, 1971: January). A two-way television network would be required to provide the viewers with a home terminal through which they could register their responses. The home terminal is a vital part of such a communication system, as it will directly affect the richness and nature of the viewers' responses.

Marshall McLuhan (1965: 18) wrote that "the effects of technology do not occur at the levels of opinions or concepts, but alter the sense of ratios or patterns of perceptions steadily and without resistance. The serious artist is the only person able to encounter technology with impunity just because he is an expert aware of the changes in sense perceptions."

The realization that technology affects its audience in a subtle, subliminal manner means that the "artist" who designs the technology and those who introduce it possess a tremendous power which could be highly political in nature. By introducing an attractive technological

AUTHOR'S NOTE: This paper was sponsored by Professor I. de Sola Pool and was supported by a grant from the Center for International Studies, Massachusetts Institute of Technology. I wish to thank Professors D. Lerner, I. de Sola Pool and T. B. Sheridan for their guidance and useful advice.

I would also like to thank Mr. Richard Miller from the Communications Department for helping to shape the ten questions at the end of the experiment, and extend my thanks to those who participated in the experiment for not only providing feedback response, but for giving a general appraisal of the experiment as a whole.

innovation the artist could shape the perceptions and behavior of his audience. A subconscious manipulation is dangerous regardless of the end. It deprives people of the right to examine the imposed change and resist it. But since technological growth and increasing cooperation between scientific disciplines are inevitable, society must deal with the question of how to minimize the risk involved in employing technology to manipulate change.

One way to minimize subconscious manipulation is to make the audience aware of the purposes and possible effects of the technological innovations. The main disadvantage of this approach is that it is difficult to perceive the "message" of the new medium beforehand or when it is first introduced. This difficulty is compounded by the fact that it is the interaction of the new medium with other media that could trigger the significant changes. Those who introduce the technology may not be conscious of the effects of the medium and may themselves be subconsciously manipulated by their media. Since there is not much we can do about what we are not conscious of, a full public disclosure of the intentions of those who introduce the medium is the best tool to reduce such manipulation.

A second way to reduce the risk of subconscious effects of technology is to analyze the medium within the context into which it is being introduced. Since it is impossible to detect all the changes which a new technological innovation introduces, this analysis has to be a continuous one. It should begin prior to the introduction of the technology and should continue indefinitely.

This paper is a first attempt to examine the relations between the possible technological features of a two-way TV home terminal, and some of the objectives offered in the literature concerning two-way TV. The analysis will be done on the sociopolitical level rather than on the level of the effects of such interaction on the viewers' perceptions.

OBJECTIVE CONSTRAINTS ANALYSIS FRAMEWORK

Analyzing the technological design of a home terminal through the examination of the broad context of two-way TV could help the designer answer such questions as: (1) How important is it to enable the listener to state specific requests? (2) Would the optimal terminal be the one which is most versatile yet the cheapest?

These questions involve economic trade-offs which depend on the weight society puts on the different goals of two-way communications. The problem of designing an optimal home terminal is analogous to a programming problem where the objective is to maximize (or minimize) a value function subject to a lot of constraints.

Here the value function we wish to maximize is the social utility of two-way communications. The operational boundaries are constrained by:

(1) technological advance (available channels for communications);

(2) human nature (attitudes toward interaction with machines, rate of information absorption, memory);

(3) economic resources; and

(4) political and social implications [see Table 1].

Utility theory is not yet sufficiently advanced to allow the transmutation of nontangible constraints and objectives into graphs whose intersections define the parameters of the optimal terminal. In order to use such a utilitarian approach, we must find some common denominator to translate all the objectives and constraints, and to determine what weights to assign each objective (priorities). The first part can be partially achieved by describing the objectives of two-way communications in terms of the technical features required. This information is necessary to determine the costs of such objectives. Establishing the weights, however, is a difficult task which should be undertaken by society, and it becomes the responsibility of science to supply society with guidelines to determine these weights. This can be achieved by exploring the benefits and the social and economic costs that could result from accomplishing the objectives.

Table 1. Major Constraints on Two-Way Communications

- C_1 Technological
  - hardware
    - available communication technologies (c_{11})
  - software
    - methods for analyzing feedback (c_{121})
    - procedures for different modes of communications (c_{122})
- C_2 Human Nature
  - physiological
    - rate of information absorption (c_{211})
    - memory (c_{212})
  - psychological
    - attitudes toward personal versus impersonal communications (c_{221})
    - emotive versus rational decision making (c_{222})
    - frustration due to lack of response (c_{223})
- C_3 Economic
  - support
    - governmental sources (c_{311})
    - public subscription (c_{312})
    - services (c_{313})
  - costs
    - research (c_{321})
    - hardware (c_{322})

Table 1. (continued)

C4 Political
- present political structure is based on limited citizen feedback (reconciliation process) (c41)
- surveillance potentials of two-way communications (c42)
- skewed use of facilities by one segment of the population (information-rich versus information-poor) (c43)
- frustration due to lack of response (c44)
- "majority tyranny" (c45)
- emotive rather than rational response (c46)

MAJOR OBJECTIVES FOR TWO-WAY COMMUNICATIONS

Table 2 lists some of the major objectives for two-way communications. These objectives can be divided into four major areas: politics, education, social, and services.

In politics, the major objective is to make government more responsive and improve its decision-making by improving the communications channels for feedback from the citizenry. The broad objective has different implications and constraints at different societal levels: group, community, and state. At each of these levels, the broad objective—making government more responsive—has different sub-objectives which are outlined in Table 2.

There are two main denominators which can be used to divide society for purposes of the political analysis: (1) use as a base the relative power of groups and their communications with the decision makers (i.e., divide society into levels of relative influence for the purpose of analyzing the effects of two-way communication on decision-making at all levels of government (see figure below):

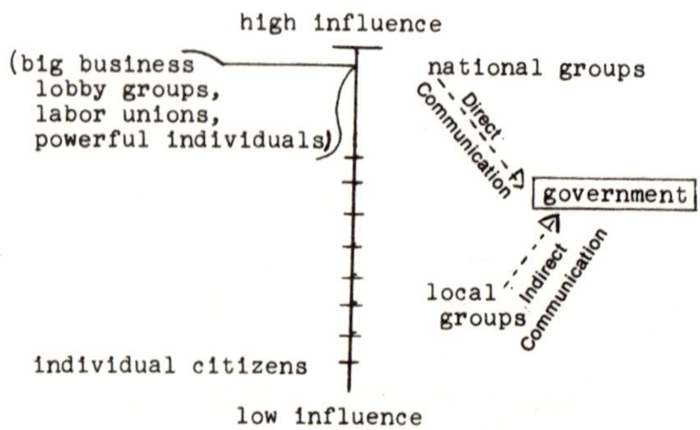

Table 2. Major Objectives for Two-Way Communications

1. Participatory Democracy
   - group level
     - help on-going groups improve their decision-making processes by increasing the individual input into the discussion (0111)
     - form new political groups (encouraging peripheral groups to take part in the decision-making) (0112)
   - community level
     - help community discuss issues of interest (0121)
     - aid community decision-making (0122)
     - enable members of community to communicate with elected officials (0123)
     - educate the community about important issues (0124)
   - state level
     - exchange of information with state leaders (0131)
     - voting on state and national issues (0132)
     - education of the public about issues of national concern (0133)
2. <u>Education</u> (individual selection)
   - information retrieval
     - getting library information on CRT
     - comparative study of library information
   - learning a skill or trade
     - simple skill
     - complex skill
   - responding to a program not selected by the individual
     - panel discussion
     - educational program
3. <u>Social</u>
   - form social groups
     - only initiation
     - CTV as main media for communication (handicapped people)
   - mating of individuals
4. <u>Services</u>
   - shopping
   - medical (Tally medicine)
   - post (sending and receiving post via CRT)
   - entertainment (ordering movies, etc.)
   - games (bridge, gambling)
   - bulletins reporting advances in specialized professions
   - computer interaction
   - wide news coverage

(2) use as a base a geographical location and community size (e.g., divide society into levels according to community size and geographical location for studying the effects of two-way communication on politics at these levels (see figure below).

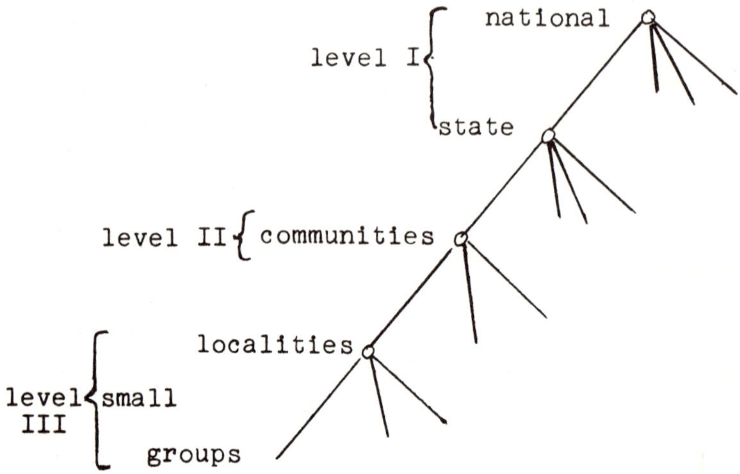

I have chosen the second division for the reason that cable TV *two-way potential* is predicted to be used for political purposes on a local level (e.g., community discussion with elected officials, local origination of programs, etc.). A division into levels according to relative influence on political decision-making should be done for the purposes of an objective-constraint analysis. Present decision-making processes are one of the major constraints on the use of cable TV for two-way communications. Any radical change in these processes could cause social instability. Such an analysis should focus on how changes in decision-making processes on the *local* level will affect present decision-making processes which are often independent of geographical locations (i.e., study how the power of an interest group such as the labor unions will be affected by changes in local decision-making processes).

A second major objective of two-way communications is education. The potential for interaction with the TV, and the new developments in man-computer interaction and computer memories, make this objective very attractive. Computer computations, books from libraries (stored in computer memories) and computer graphics (making actual design on the TV screen) can be transmitted to the home TV. Individual self-teaching of skills and trades can be made possible.

In Table 2, the objectives of education were selected according to presently developing research areas.

A third major objective of two-way communications is to improve social relations within a community. The selection of the objectives is based on my own predictions.

The fourth major objective of two-way communications, which will probably be the main reason for its introduction, is services. The objectives listed in Table 2 are partially based on prediction of which services will yield most profit (entertainment, shopping, games) and others are predicted according to presently developing research (medical, computer interaction).

TECHNOLOGIES AVAILABLE FOR TWO-WAY COMMUNICATIONS BETWEEN HOME AND THE STUDIO

Table 3a contains a description of the technologies available for transmitting information from the broadcasting centers to the home reserves and in the opposite direction. Table 3b describes some of the features which are mentioned in Table 3a in more detail. For obtaining most of the two-way communications objectives, a combination of these technological features will be needed. For example, present one-way TV broadcasting is a combination of boxes 2/1 and 3/1a (studio and to home audio and video).

Table 3a enables an evaluation of the versatility of presently developing home terminals in terms of the communications objectives. The technological evaluation can be done in two steps: (1) once an objective is selected, the table can help establish the necessary technological features needed to achieve this objective; (2) the home terminal can then be evaluated on the basis of the technological requirements (to what extent these requirements are satisfied).

Following are a few examples which demonstrate how to interpret the boxes in Table 3a:

Box 2: The only transmission channels available are one-way audio communication from the studio to the home. Two channels can be used for this purpose: (1) the telephone; (2) the TV set.

Box 11: The only transmission available is one-way video from home to the studio. There are two alternatives for this transmission: (1) use of home light camera for live transmission (examination by a doctor); (2) use of video tape for indirect transmission (home origination of a program). See Table 4 for actual use of this table.

Table 3a. Possible Technological Channels for Communications*

A — STUDIO TO HOME (down stream)

HOME TO STUDIO upstream	B	Ø	one-way audio	one-way audio	one-way print	one-way signal
	Ø	1) no communication	2) a) voice through TV b) voice through telephone	3) a) continuous picture-- 1) individual 2) public channel b) stationary 1) channel sharing (frame grabbing)	4) a) xerox copy via cable b) strip printers c) screen photos d) regular mail	5) a) simple--numbers and video messages b) complex--computer graphics
	one-way audio	6) a) telephone b) microphone in home terminal	7) H→S audio S→H audio	8) H→S audio S→H **video**	9) H→S audio S→H print	10) H→S audio S→H signal
	one-way video	11) 1) home camera a) direct transmission b) indirect transmission (video tape)	12) H→S video S→H audio	13) H→S video S→H video	14) H→S video S→H print	15) H→S video S→H signal
	one-way print	16) 1) signal sent from home is interpreted by a computer and printed as computer output	17) H→S print S→H audio	18) H→S print S→H video	19) H→S print S→H print	20) H→S print S→H signal
	one-way signal	21) **S) Simple--less than 12 buttons to respond with** **C) Complex--use of entire screen for feedback**	22) H→S signal S→H audio	23) H→S signal S→H video	24) H→S signal S→H print	25) H→S signal S→H signal

Key: Ø = no communications S = Studio H= Home *please see guide in following page

Table 3b.

	Technological Features	*Description*
3a	Continuous picture, individual	Each home receives a private TV channel which is connected to a program exchange. (REDIFFUSION SYSTEM).
3b	"Frame grabbing" technique	30 full TV pictures are transmitted per second. A small video tape located in the home terminal can record one full picture in 1/30 of a second and replay it on the screen. If a single picture is shown for 10 seconds, 600 people can get individual pictures using only one channel. A computer is used to locate the different requests (MITRE SYSTEM).
4a	Xerox copy via telephone	Xerox company has developed a device which allows sending Xerox copies via telephone lines. This capability can be added to the home terminal (newspapers, printed information can be sent via cable in seconds).
4b	Strip Printers	100 words a minute. Printers are available which print messages on strips of paper.
4c	Screen Photos	Polaroid camera can be used to make "copies" from screen displays.
5b	Complex computer graphics	TV screen connected to a computer can be used for design using a "light pen."
21c	Use of entire screen for feedback	The entire screen can be used for feedback. Two techniques are available for this purpose: light pen and photo-electric light beams. The first will allow feedback by touching the screen with a finger (a design using these techniques was completed at the end of January 1972).

Table 4. Breakdown of Objectives into Technological Components

Objective *Participatory Democracy*	*Technological* *Features Needed**
Group level:	(boxes in Table 3)
1) group discussion	$2^a6^a11^b21^{bs}$
2) forming action groups	$2^a3^a7^a13^c21^{bs}$
Community level:	
1) community discussion issues	$2^a3^a7\ 21^{ac}$
2) community decision-making	$2^a3^a7^b21^{ac}$
3) community education	$2^a3^a7\ 21^{ac}$
State level:	
1) respond to a speech	$2^a3^a21^{ac}$
2) voting	$2^a3^a21^{as}$
3) education	$2^a3^a21^{ac}$
Education	
Information retrieval:	
1) obtaining library information	$3^a4^c7^c21^{as}$
2) analysis of information	$3^a4^c7^c21^{ac}$
Learning a skill:	
1) complex skill	$2^b3^a4^b21^{ac}16^c$
2) simple skill	$2^b3^a4^b21^{as}$
Responding to a program:	
1) panel discussion	$2^a3^a6^a21^{ac}$
2) educational program	$2^a3^a6^c21^{as}$
Objective	*Channels Needed*
Services:	
1) shopping	$2^c3^a6^b13^c21^{as}$ or ^{c}
2) Tally medicine	$7^a13^a21^{ac}$
3) post	$4^{al}21^{ac}$
4) entertainment	$2^a3^a21^{as}$
5) computing	$3^a4^{a2}5^{a2}21^{ac}25$
6) games — with others	13^a7^a
with studio	$2^a3^a21^{as}$or ^{c}

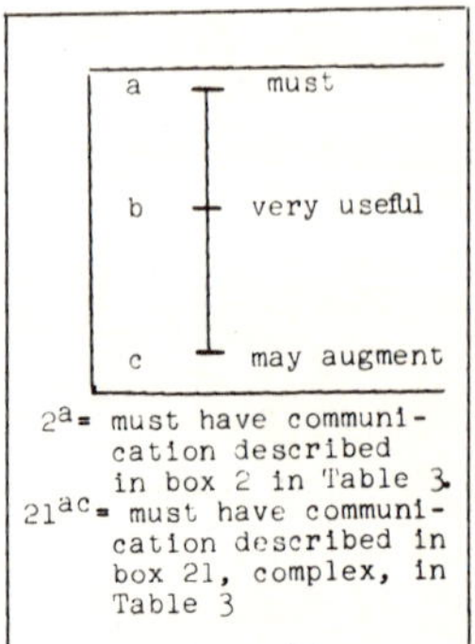

Key for table

* The choice of these technological features is based on a prediction as to how these objectives will be achieved. A final choice of these features will depend on utility-cost tradeoffs and experimentation.

ANALYSIS

A cross-variable analysis of the objectives and constraints of two-way communications could be very useful in expanding our knowledge of the applications and limitations of two-way communications. This analysis will uncover many areas for research such as the effect of attitude toward two-way nonpersonal communications on willingness of humans to use it.

Tables 1 and 2 list some of the important objectives and constraints of two-way communications. Table 3 describes the complete set of communications channels between the home and the broadcasting center. Table 4 summarizes the breakdown of the objectives into the communication channels required to achieve them.

A cross-variable analysis of Tables 1 and 2 will provide the guidelines for determining the weights to be assigned to the different objectives. Tables 3 and 4 provide the information for establishing guidelines describing specific technological features needed in order to obtain a certain objective, the first step toward determining its costs.

Table 5 is the cross-variable matrix of the constraints and the political objectives. In this paper, only the shaded areas will be discussed. (The design variables which relate to the human nature and economic constraints having an effect on both the complexity of the terminal and the possible financial sources available for support will not be discussed since both relate to a political use of two-way communication, and thus would necessitate too broad a task.) A complete analysis of all the objective-constraint cross-variables would involve such disciplines as psychology, sociology, and economics. Tables 1 and 2 offer researchers from the different disciplines some guidance as to which variables should be included in a cross-variable analysis. To construct a cross-variable analysis, one has to select the issues of interest from the tables of objectives and construct a table similar to Table 5 for developing a cross-variable matrix.

USE OF TWO-WAY COMMUNICATIONS FOR "PARTICIPATORY DEMOCRACY"

Two-way communications can be applied in three societal levels: group, community and state-national. The objective of such citizen feedback is to increase the individual input into the political decision-making processes and make the citizen more knowledgeable about society.

On a group level, the objectives of two-way communications could be the following: (1) help ongoing groups improve their decision-making

Table 5. Cross tabulation–Two-Way Communication Constraints and Political Objectives

OBJECTIVES

CONSTRAINTS		o111	0112	0121	0122	0123	0124	0131	0132	0133
TECHNOLOGY	C111									
	C121		A			C			E	
	C122									
Human Nature	C211									
	C212									
	C221									
	C221									
	C222									
	C223									
ECONOMIC	C311									
	C312									
	C313									
	C321									
	C322									
POLITICAL	C41									
	C42									
	C43		B			D			F	
	C44									
	C45									
	C46									
		GROUP			COMMUNITY				STATE-NATION	

processes by increasing the individual input into the discussion; (2) form new political groups, making access to the media easier and thus encouraging peripheral groups to take part in decision-making (e.g., neighborhood political groups, PTA, welfare mothers, etc.).

On the community level, the objectives of two-way communications could be the following: (1) help the community discuss issues of interest by enabling citizens to bring issues to the attention of the community and educate its members on these issues; (2) aid community decision-making by having a series of discussions about some issues of concern followed by voting–analogous to town meetings; (3) enable the members of a community to communicate with the elected officials (school board, city council, congressmen), educate the community about important issues (giving the background and rationale for decisions regarding welfare, zoning rules, taxes, etc.).

On state-national level, two-way communications could be used to exchange information with state leaders (congressmen, senators, executive branch); to vote on state and national issues; educating the public about issues of state or national concern (desegregation, senate hearings, etc.).

Following is the discussion of the shaded areas in Table 5.

A. Group Objectives and Technological* Constraints

The main technological constraint on achieving the group level goals pertains to software characteristics. For an effective small group discussion direct face-to-face interaction is necessary. Members of the group should be able to express verbally their opinions and be able to affect the direction of the discussion. It is possible, technologically, to carry such group discussions via cable TV, but the expense of the technological features (see Table 4) may be prohibitive.** The limited communications channels of studio to home audio and video, and home to studio audio or digital feedback (using buttons) can be used by cable TV both to spawn and measure an interest group's reaction to an issue. Community centers could be established for group meetings. In these centers (special studios designed for community use) a limited two-way feedback mechanism can supplement the group discussion.

B. Group Objectives and Political Constraints

The major major political constraint on group objectives (helping ongoing groups improve their decision-making processes and form new interest groups) is that the present political process is based on limited citizen feedback from both direct and indirect sources. Most of the direct feedback comes from organized interest groups (such as business, labor unions, religious groups). The indirect feedback comes from the population through communication channels such as elections, public polling, social unrest, talk shows, etc. The use of cable TV for the purpose of

---

**Technological constraint refers to both software and hardware considerations.*

***Whether these features are desirable is discussed in the preceding cross-variable analysis.*

forming new interest groups could be expected to extend the socio-economic bounds in the social pyramid from which pressures will be exerted on the political structure. The groups which presently control the decision-making powers may feel threatened by this process, thus leading to social tensions or social repression. If the present political system is not flexible enough to allow new groups to share in the decision-making, social instability may result. Frustration of these new groups due to the lack of positive response from the system could sway them to social apathy or social outbreaks. Both are dangerous.

An additional political constraint is a skewed use of the facilities by a few dominant groups. This could widen the gap between ongoing powerful political groups and groups that do not have equal access to the facilities. Legislative channels could properly deal with this constraint.

In deciding what weights to assign the group objectives, society must be aware of these constraints and provide the proper guards to prevent social instability and misuse if a decision is made to use CTV for achieving the goals on the group level.

C. Community Objectives and Technological Constraints

The community objectives listed in Table 2 involve discussion of issues between a large group of people and few decision makers, speakers, or discussion moderators. As in the group level discussions, the main technological constraints lie in the software area. There are no techniques as yet that allow instant analysis of mass-verbal comments. An additional constraint develops as all the members of the community make verbal input to the speakers–no complex discussion can ensue. Guidelines to avoid such disturbances yet allow meaningful group feedback must be developed (see Appendix A). The computer technology can be a useful tool in analyzing a nonverbal digital feedback. The use of the computer will enable instant analysis of the group feedback, an essential for a meaningful feedback during a discussion. It is important to allow group response during the entire discussion and not only at the request of the moderator. Limiting the community feedback to instances when the moderator decides to pool opinions will lower the validity of feedback for the following reasons:

(1) the moderator can limit the feedback to issues of interest to him;

(2) there will be no data at the community level of understanding of the issues;

(3) the community can have very limited effect on the discussion; and

(4) no meaningful education of the community about issues will be possible without enabling the listeners to elicit such responses as "do not understand," "give more information," or "move to another topic." The listeners should be able to elicit these responses during the entire discussion.

Without these capabilities it is doubtful that cable TV two-way potential can be used effectively for political purposes.

Table 4 shows that for the four objectives on the community level, a complex one-way (home to studio) signal transmission will be needed to allow for the feedback discussed above.

Another software constraint is the absence of a discussion procedure for the different communications objectives. These procedures (meshing technological capacity with discussion goals) are essential for the successful use of two-way communications for political purposes.

An additional constraint is the lack of statistical tools that will enable a community and its elected officials to measure the degree to which a given feedback represents the community sentiment. The scientific method used in public polling cannot be directly applied to such citizen feedback due to the biases of the group participating in the discussion. It is reasonable to expect that the development of such statistical tools should be more reliable as feedback through cable TV becomes more widespread.

It is technologically feasible to build a system which will allow the complex one-way signal transmission necessary to achieve objectives on the community level.

D. Community Objectives and Political Constraints

The technological feasibility of two-way communications on the community level is shaded by the political constraints. The commercial attractiveness of two-way communications for commercial purposes may make the technological development inevitable (see Table 4). The scientists who develop these channels must be aware of the possible social consequences. The scientist's intentions may not necessarily parallel the guidelines of the politician who decides to use these channels for political purposes. It is therefore very important to project possible political implications of a technology which seems at first glance apolitical, yet may result in social instability if proper measures to prevent its evils are not considered in advance.

As at the group level, the main political constraint is that the present political system is based on limited citizen feedback. Organizing communities around issues will increase feedback such that failing modification of the present decision-making processes which incorporate feedback, social instability is inevitable. This increased feedback may limit the decision-making powers of the community-elected officials (or increase their power if they are able to manipulate the system) making reconciliation of issues more difficult. Some scientists (Heinz Eulau, 1970: 187) see such citizen participation as a "participatory nightmare." For a political system whose decision-making methodology is based on the reconciliation processes, this is a real threat. Although town meetings demonstrate the plausibility of community decision-making, town meetings take place in relatively few communities and involve face-to-face interactions. Research must be done to answer such questions as: (1) how will impersonal communications affect the community voting behavior? (2) will the technological channels improve communications between people (in similar situations) rather than restrain it? (3) can findings from a study of few communities be projected to other communities? (4) how effective is the town meeting decision-making process in comparison to city council decision-making?

A second constraint is frustration due to lack of response. Assuming proper statistical tools are developed for this purpose, the communities will have, for the first time, a clear idea of what they want. A lack of response from the elected official may stimulate social instability. The "silent majority" support often claimed by politicians will no longer be "silent"; the politician will be forced to act according to the will of the community, a frequent impossibility. He may be forced to explain and educate the community on his decision-making rationale, but even then, the threat of social instability is real.

A third constraint is that as participation in community decision-making becomes as effortless as pressing some buttons at home, people will react emotionally rather than rationally on issues about which they have insufficient information. At the present, some effort and time from citizens to input feedback into the political system is needed. The people who are willing to give this feedback are usually directly affected by the issues and consequently are more informed.

A fourth constraint involves the concept of "majority tyranny." The democratic concept calls for majority decisions. But is it fair that the votes of those individuals most directly affected by a decision should have an equal weight as the votes of the rest of the community? In the present political system, it is easier to protect such a minority as "majority wants" are rarely clearly defined. Introducing citizen feedback on the community level may make this conceptual problem very real.

Even though the constraints are severe and the dangers real, there is much to gain by the use of cable TV on the community level especially as a tool to educate people on issues. This justifies research toward achieving the community goals. We must be aware that legislative means may be insufficient to deal with all the political constraints since no law can force an individual to react rationally rather than emotionally. Guidelines and procedures for decision-making must be developed to avoid problems that could arise from emotive and ill-informed decisions.

E. Societal-National Objectives and Technological Constraints

The type of communications channels that are needed at this level of communications are similar to those needed on the community level (see Table 4). Similar technological constraints apply to both levels and are discussed in the preceding section.

F. Societal-National Objectives and Political Constraints

The nature of the political constraints makes it prohibitive to experiment with two-way communications on the societal level before a thorough examination of all the problems associated with the use of two-way communications on the group and community levels. Even when all problems associated with the community and group levels are satisfactorily resolved, a careful consideration must be given to the validity of projections from these levels to broader societal levels. Using citizen feedback on a national level may require a fundamental restructuring of the entire democratic structure—a huge task indeed. Research in this area is called for to expose the dangers involved, to insure the politicians' awareness of these dangers.

The political constraints on the societal level are similar to those on the community level—a level on which it is difficult and dangerous to experiment. Even when we develop satisfactory knowledge of what happens on the community level, it will be difficult to predict what effects the use of two-way communication at the community level will have on the political structure. These effects could make the use of two-way communications on a national level more feasible or more prohibitive than is now apparent.

CONCLUSIONS

The purpose of this paper was to link technological design considerations of a home terminal with the possible political consequences of such design. It was demonstrated that a home terminal may be versatile, may be economically and technologically feasible, and still may not be desirable from a societal standpoint. A versatile home terminal could promote the development, by naive designers, of two-way communications which could threaten the stability of society. Once the technological features are available on a mass level, the social scientists aware of the dangers may not have the power to limit their use.

APPENDIX

INTRODUCTION

The process of absorbing information that begins every day when we wake up can become very frustrating without a meaningful outlet of response to that information. We absorb data about the world around us and about issues that affect us directly, but usually the most we can do is communicate our thoughts to a close circle of friends who in most cases do not govern the decisions which concern us. Many people allow much of this information to become a part of their lives without talking about it even with close friends. They become almost hypnotized by a monologue with themselves. This process can become more and more frustrating as the amount of information with which individuals are bombarded increases.

A group of scientists with heightened social awareness is beginning to realize the potentials of the existing technologies in furnishing people with channels to communicate their feelings and thoughts to others. The potentials of cable TV and the possible establishment of two-way communications between the citizen and the media make this type of communication imminent.

There are three main objectives to this study: a preliminary design for a home terminal which would enable the listener (viewer) to respond to a panel discussion which takes place in the broadcasting studio, an investigation of the way both the speakers and the listeners can benefit from such a flow of information; and a study of how people view participation in such a discussion when they must channel their responses into a certain number of preset options.

In designing the imaginary terminal my purpose was to enable the listeners to express opinions on specific ideas voiced by the speakers and to ask the speakers to respond to specific ideas expressed by other speakers, or to further explain some of the listeners' own ideas. The system will also allow the listeners to present questions to other listeners.

A distinction should be made between the design of a terminal which asks the audience for their attitudinal response, as in the terminal discussed above, and the design of a terminal which enables people to participate in actual decision-making policies. The study of the latter terminal is discussed by Professor T.B. Sheridan (January 1971). In order to fulfill these objectives, the participants were asked to visualize themselves in a "live" situation and to imagine that their responses will have some effect on the discussion. The fact that their participation is imaginary limits the validity of the results obtained. As one of the participants expressed, "In a live situation I probably would have had fewer responses, but they would have been stronger." The information that can be gained by this study could prove very useful for a future design of real-life experiments, although the absence of the visual dimension of the TV experience is a significant loss.

THE PARTICIPANTS

The participants (about 10), mostly professors and doctoral students at Stanford University, were chosen from the following fields: Sociology, English, Engineering Economic Systems, Industrial Engineering, Business, Communications and City Planning.

DESCRIPTION OF THE EXPERIMENT

The participants were asked to visualize themselves in a "live" situation: "There is a discussion on your home TV about student revolt (taken from *Playboy Magazine* 1969). . . . Thanks to technology, besides your TV set there is a terminal through which you can make some responses to the discussion . . . . It is important that while 'listening' you imagine yourself in a live situation where your responses will have some effect on the discussion."

On the terminal there was an additional screen on which the words of the speakers appeared "simultaneously" with their imagined voices. In other words, in this experiment all that the participants saw was the typed discussion—not the speakers themselves, and they were asked to imagine themselves in a situation where they could see the speakers.

The terminal screen was labeled vertically with letters that corresponded to lines on the screens in order to enable the "listeners" to locate their responses to the idea expressed in that line. On the terminal were coded responses (see Figure 3) and the participants were asked to choose among them by pressing (making a check on the button) the button that corresponds to the response of their choice.

A questionnaire was included at the end of the discussion in which the participants were asked to evaluate the system both as listeners and as though they were the speakers.

In a real life experiment, listeners (viewers) would be required to switch their attention from the speakers to the printed letters which could be more disturbing than the situation in which they responded to the

Fig. One
Terminal Used in This Experiment

Fig. Two
An Alternative Terminal

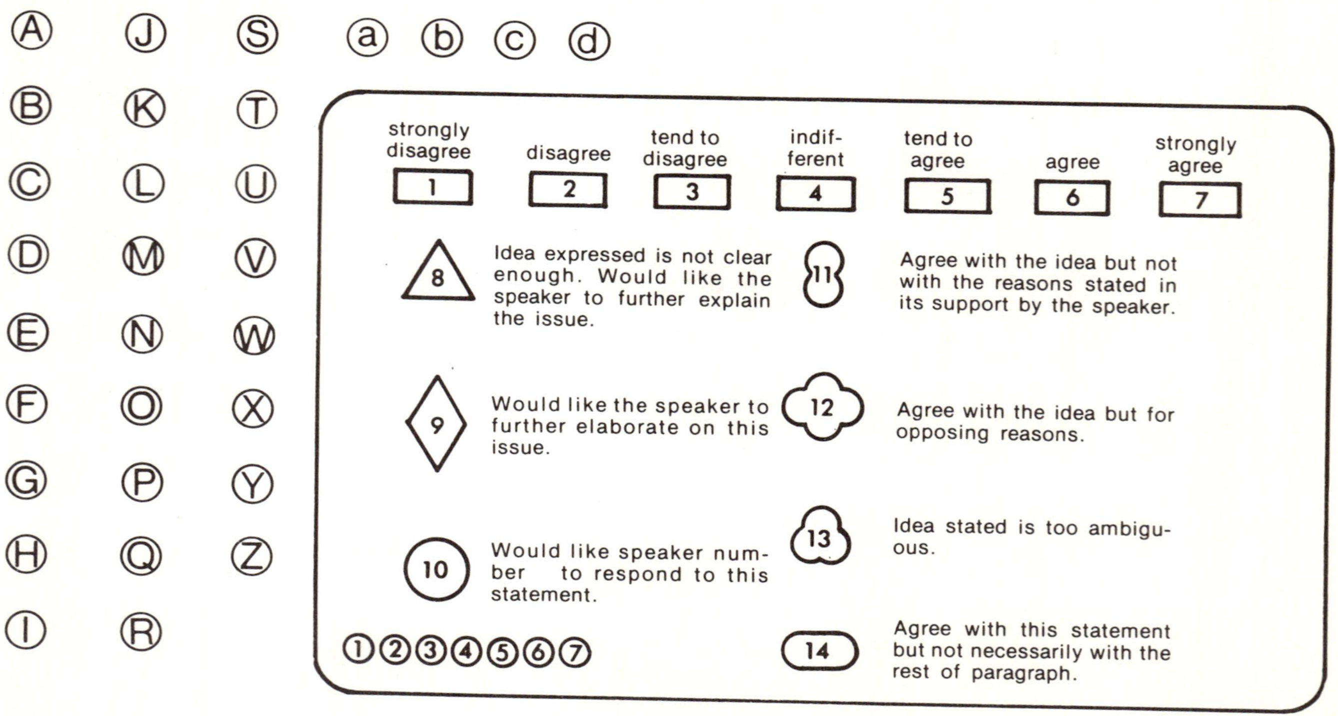
A B C D E F G H I
J K L M N O P Q R
S T U V W X Y Z
a b c d
strongly disagree 1
disagree 2
tend to disagree 3
indif-ferent 4
tend to agree 5
agree 6
strongly agree 7
8
Idea expressed is not clear enough. Would like the speaker to further explain the issue.
11
Agree with the idea but not with the reasons stated in its support by the speaker.
9
Would like the speaker to further elaborate on this issue.
12
Agree with the idea but for opposing reasons.
10
Would like speaker number to respond to this statement.
13
Idea stated is too ambiguous.
1 2 3 4 5 6 7
14
Agree with this statement but not necessarily with the rest of paragraph.

Fig. Three

discussion in the experiment. The results of this experiment showed that the participants did not respond often enough to disturb their "listening" (reading).

THE DEVICE

In order for the viewer to be able to make responses to some specific ideas during the discussion, not only during periodic intervals determined by the discussion moderator, the discussion should appear in print in front of him. This can be done by simultaneously printing what is being said on stage and transmitting it to home TV screens. In this experiment the screen was a part of the terminal (Figure 1), but it will be less costly to use a portion of the TV screen on which the discussion is being displayed for this purpose (Figure 2).

As can be seen in Figure 2, the lower right portion of the screen can be reserved for displaying the feedback statistics. The size of the letters to be displayed and how many lines shall be displayed at a time is a topic for further study. In the experiment, the participants were allowed to refer back to previous pages before they made a response, but were asked to indicate if they did so in the questionnaire. Some indicated that they referred back once or twice, one indicated he did not refer back, and the others did not indicate anything. Clearly, no valid conclusion can be drawn from this data with regard to the number of lines to be displayed but it is encouraging that no one had indicated that it was to his advantage to have the entire discussion in front of him for reference, a situation which will not exist in reality. It can be expected that the number of times the participants will refer back to previous lines will vary according to the interest he has in what is being said and the complexity of the discussion.

In order to be able to respond to a specific idea which is expressed in a certain line of the printed screen, vertical letters are assigned to each line and the participants can localize their response by pressing the button marked with the letter that corresponds to that line. The selection of the response will be done in the same way. These responses will be stored in a computer memory.

Figure 3 describes the responses available to the listener. These responses were carefully designed so as to avoid distortion of the listeners' opinions and to enable them to affect the discussion and its content in some limited way. Of course, there is no optimum set of responses that could enable the listeners to express their opinions free of distortion, and the object of the research is to investigate how to minimize these distortions and study their possible effects on the content of the discussion. This is one of the real dangers of the use of preset responses and a topic which should be studied continuously.

DISCUSSION AND ANALYSIS OF THE RESULTS

The following is a discussion of the participants' responses to the questionnaire. First, a question will be presented and then a description of the participants' answers to this question which will be followed by a brief analysis of these answers.

THE DEVICE

Question 1: Did you want to express or elicit any reaction that you could not by using the terminal? The general criticism of the responses was that only rational responses could be elicited and that there were none to express emotive responses such as "rubbish," "naive," or "dislike" (vs. disagree). One participant wanted to agree with a statement and its supporting comments but not with the conclusion and another wanted a button to express "good example" or "good analogy." Yet another found it frustrating to simply mark agree or not: "these were the reasons that were compelling, agree and disagree imply a quantitative measure along a given dimension." But according to this participant, agreement and disagreement are usually along different dimensions. One participant who used every space available to respond, did not want to add anything.

Analysis: The frustration that some expressed in not being able to elicit a response that would best describe their feelings and opinions will always occur when preset responses are being used. Fortunately, no electronic media can replace the face-to-face interaction in which the communications take place on many different levels. The purpose of the preset response terminal is to create a feedback channel not yet in existence. The frustration people have in responding through such channels of preset responses may induce them to seek other ways to communicate.

An encouraging result was that one of the participants indicated that initially she resented the idea of responding to a "robot" but later found it useful to be able to respond to the discussion via the terminal.

Figure 4 shows the number of times a specific response was used by the participants. This distribution can aid in evaluating the responses.

It must be kept in mind that the choice of some specific responses over others is clearly a function of the interest the listeners have in the discussion, and the terminal should be tried many times with different topics of discussion before any conclusion can be drawn. What can be learned from Figure 4 is that the responses were used such and such number of times. Such information can be useful in adding or deleting some responses.

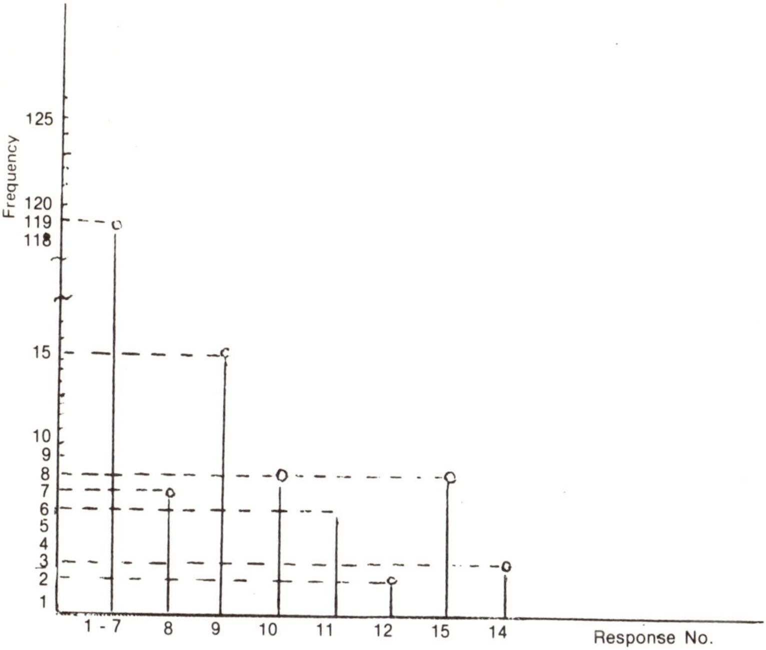

Fig. Four

Question 2: Were your responses positive? Negative? Equal? Figure 5 shows that the participants reacted equally in all these categories.

Analysis: Some people responded when they agreed, some when they disagreed and others responded both when they agreed or disagreed. Knowledge of such information is important in the design of a response terminal. Ideally, a terminal should allow responses that suit all the above, but due to the limited number of responses which can be handled effectively, an emphasis could be put on designing the response to gravitate toward the positive or negative if it is clearly shown that people lean toward one or the other.

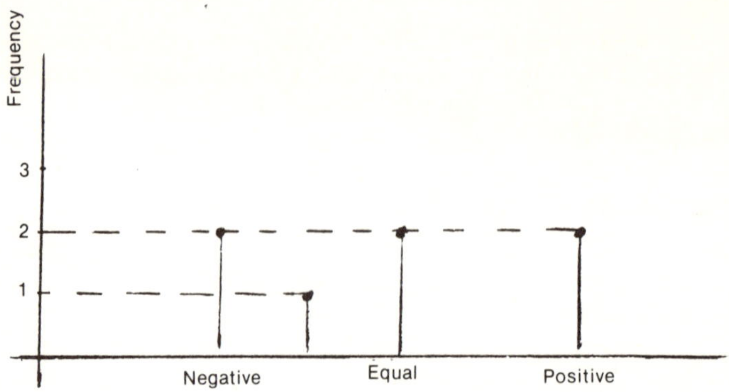

Fig. Five

Question 8: What other general preset panel responses would you want on this instrument? The responses of the participants to this question were similar to their responses to question 1. The intention in the question was to get some specific responses the participants wanted to add but their answers to this question were in more general terms.

According to the participants, additional responses should allow for the following: measure emotional responses to attitudes and to language, indicate level of general interest or boredom, press panelists to move onto another topic. One participant expressed a wish to be able to "type" questions or comments on the letter part of the keyboard but the cost of such an improvement would be prohibitive. (An alternative for this suggestion will be discussed in the Conclusion.)

THE LISTENER

Some of the questions in the questionnaire were directed toward the participants as listeners to the discussion and in some questions the participants had to assume the role of the panelists. In this section those questions that relate to the listener will be discussed.

Question 3: Do you think that a panel such as this would be useful, a hindrance, or make no difference for an audience listening to a panel discussion? The reactions of the participants to this question ranged from "very useful" to "hindrance." Some felt that the discussion would not be able to proceed if it would be constantly interrupted by the audience. One participant thought that in a live situation it would be difficult to simultaneously listen and read the discussion. Another felt that only extreme views would be expressed. The majority of the participants did favor the use of the home terminal for such a purpose.

Analysis: The problem of interference with the discussion if the audience were to constantly bombard the panelists with responses can be solved by assigning the discussion moderator the task to set thresholds for the number of responses per idea or request, so that only after the number of responses with regard to a specific idea reaches a certain level will the moderator make it known to the panelists or the listeners, thus avoiding a constant disturbance of the discussion (guidelines for using these statistics will be discussed in the Conclusion section).

The difficulty of simultaneously watching and reading the discussion, I expect, will be minimized if the listener refers to the print only when he wants to elicit a reaction, which, as was shown in this experiment, did not happen often during the discussion. The fact that what is being said is being simultaneously printed would make it easier for the listener to locate the idea in the print.

Question 4: As a listener, would you like to see the statistics as they come in? The responses to this question range from "yes" to "not really; I would like to see how I myself respond to arguments without knowing how many people favored a particular position. At the end of the discussion some viewer statistics may be interesting." One participant would like to see the statistics in a simplified form and another saw advantage either way: "Seeing the statistics could influence me in my own responses and I might consequently not discover as fully what I feel. But seeing them as they come in would make the experience much more of a group experience."

Yet another participant would have liked structural questions to be presented to the audience at some periodic moments to prevent the disturbance of both the listeners and the speakers. The majority did favor the presentation of the statistics in some constrained and simple form during the discussion.

THE PANEL PARTICIPANT

Question 5: How would you react to the listeners' response information if you were a member of a panel discussion? Would it annoy or bother you in any way? Would you feel too much social pressure if the response information showed that the audience was strongly weighed against you? Most of the participants felt that the audience feedback would affect their thinking. "I don't see how it could but affect me just as facial cues affect me in conversation with another person, but don't think they would annoy me. . . . If I felt that the audience represented a part of the population that I respected, their strong disapproval would affect me."

Another participant felt that it would be "a bother and much harder to react to, than a live audience's boos or applause because of its cold statistical nature." Yet another felt that "I don't think I would feel social pressure–computerized information is too impersonal for that." Some participants indicated that they would make use of this information to change their argument (with honesty) in order to gain support.

Analysis: In order to avoid disturbance of the discussion, the moderator should present to the speakers only those responses in which the audience made some specific requests to the speakers (i.e., to further explain an issue). The responses on the "strongly agree" to "strongly disagree" scale should be reserved to the end of the discussion in order to avoid possible pressures on the panelists to hide their real opinions for gaining public support. The use of the terminal in this manner will help to clarify the issues by the speakers without pressuring them to modify their opinions to "fit" their audience.

Question 6: As a panel participant, would you like to see the statistics as they come in? Should there be a moderator or intermediary of some sort who informs you of responses after they reach a certain preset level? All participants favored the idea of the intermediary that would moderate the communications between the audience and the speakers. Some would like to see the statistics only at the end of the discussion. "I would be curious to see them but I might rather wait until after the panel was over unless the procedure becomes so commonplace that I responded to it as I respond to the signals in any group in which I am talking." Others would like to see them as they come in but with thresholds set high enough so interruptions would be few. One participant indicated that he would like to see the statistics only if they relate to certain specific ideas, and the main role for the moderator would be "to shape questions for the audience as the discussion evolves."

SUMMARY AND CONCLUSIONS

Most of the participants in the experiment thought a home terminal with preset responses was useful. The majority of the problems raised by

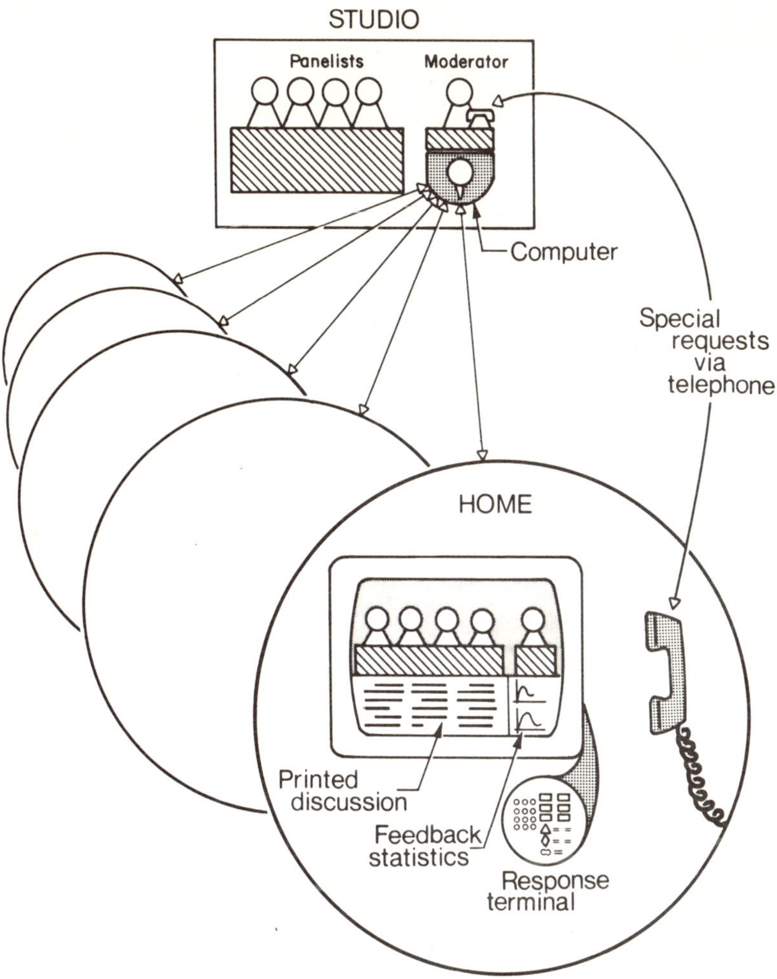

Fig. Six
The Communications System

the participants with regard to the actual use of the terminal could be solved by a careful design of the information flow. Following are some guidelines for such a communication system in light of the results obtained in this pilot experiment.

Figure 1 describes the general system. The proposed terminal is presented in Figure 2. It has the following advantages over the terminal presented in Figure 1: (1) the viewers do not have to switch their attention from the screen, except for the selection of a response on the terminal and, (2) there is no cost of an additional screen.

The presentation of the statistics to the listeners will be made on the lower right-hand part of the screen which could be blocked from view in case the listener did not want to be disturbed by the statistics during the discussion. Additional responses will be added to the terminal in order to enable emotive responses such as: "I dislike what the speaker said," "this is naive," "the discussion is boring," or, "I would like the speakers to move to another topic," etc. A button will be added to the terminal which allows the listener to respond to a whole paragraph not just to a certain line. For example, in Figure 7 (below), the listener strongly agreed with the entire paragraph printed in lines ABCD.

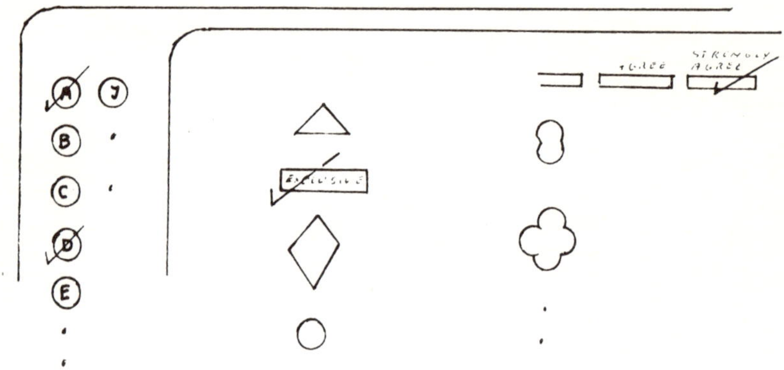

Fig. Seven

A button marked "present statistics with regard to" will enable the listeners to ask the moderator to present the feedback statistics with regard to a specific idea. An additional improvement could be the use of a *light pen* by the viewer to indicate his responses. In this case the terminal itself would be projected on the screen which would save the cost of building the home terminal (see Figure 8).

A comparison between the different alternative terminals should be made on the basis of technological feasibility and costs. If the light pen were used, there would be no need for the letter to localize an address as it could be done by touching with the pen on the specific printed line and then responding to it. There could be a great many uses for a light pen TV system (i.e., in advertising) that would reduce the costs of implementing such a system.

The selection of which feedback statistics are appropriate to present the listeners and which to the speakers, and the timing of these presentations, is of great importance. Wrong use of such statistics can disturb

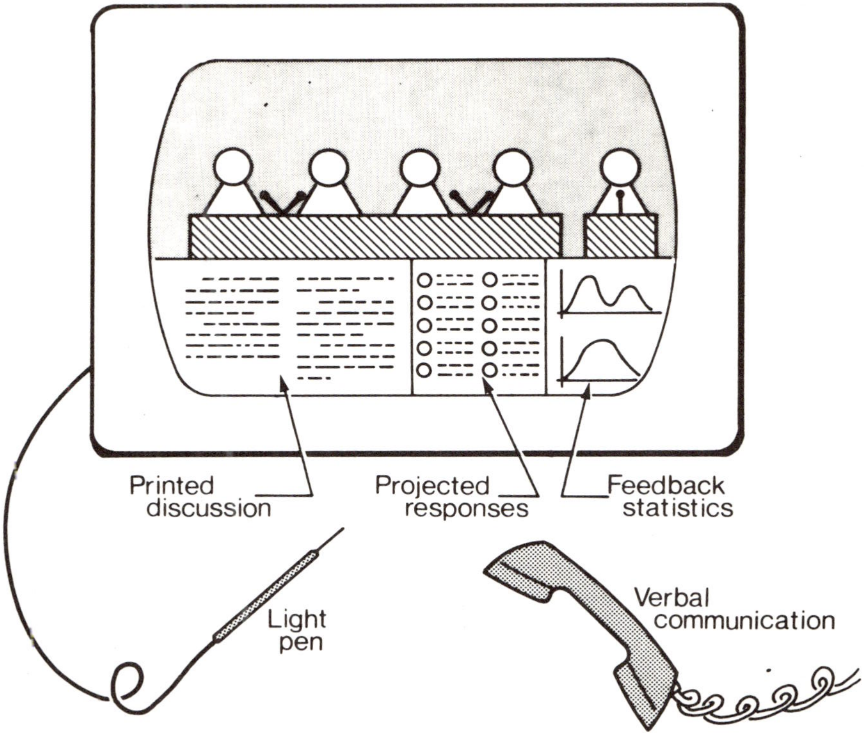

Fig. Eight
The Projected Responses

both the listeners and the speakers and even distort the content of the discussion which will result in complete defeat of the purposes of this communication.

To avoid such disturbances, the following guidelines should be adhered to:

(1) The speakers should be presented during the discussion only with specific requests from the audience (i.e., a request to respond to another speaker or to further explain an issue), and the statistics which indicate public opinion (i.e., agreement or disagreement) should be kept to the end of the discussion. The reason for this is to avoid the possibility that some of the speakers may be tempted to "fit" their argument to the audience rather than express what they really think at the cost of losing the listeners' support.

(2) The frequency of presenting the feedback statistics to the audience should be low enough not to distract the listeners' attention from the discussion. These statistics should be simple in their presentation so that no previous knowledge of statistics will be required to interpret them.

(3) Not all requests will be presented to the speakers during the discussion. It will be the function of the moderator to set the threshold high enough in order to avoid a frequent disturbance of the discussion.

(4) During break periods the moderator can present questions to the audience at the request of the speakers or by his own initiations. Members of the audience can telephone from their homes to present questions to other listeners at the end of the discussion.

(5) All feedback statistics will be shown at the end of the discussion.

The kind of communication system discussed here allows interchange not only between panel members but between various speakers and audience members, and between audience members themselves. Instead of one-way communication, with its inevitable frustrations, this system encourages a web of mutual response that makes communication dynamic rather than static. It should not only increase the social awareness of both panelists and audience but it should also make every participant more keenly aware of his responsibilities to the people with whom he is in dialogue.

References

DE SOLA POOL, I. (1971) "Public opinion." Draft prepared for the forthcoming book, Handbook of Communications, published by Rand McNally (n.d.).

———(1968) "Social trends," Science and Technology, 76 (April).

DUNN, D.A. (1971) Cable Television Delivery of Educational Services. Presented to the IEEE Eascon Conference, Washington (October).

EULAU, H. (1970) "Some potential effects of information utility on political decision makers," in Harold Sackman and Normal Nie (eds.) The Information Utility and Social Change. Montvale, New Jersey: AFIPS Press.

LEMELSHSTRICH, N. (1971) "Design of panel discussion procedures using audience feedback technology." Unpublished report prepared for D.A. Dunn, Engineering Economic Systems, Stanford University, June.

MC LUHAN, M. (1965) Understanding Media: The Extensions of Man. New York: McGraw-Hill paperbacks.

PARKER, E.B. (1972) "On-line polling and voting." July 1971 draft for the subsequent book, Planning Community Information Utilities, edited by Sackman and Boehm, AFIPS Press.

———(1970) "Technological change and the mass media," Center for Advanced Study in the Behavioral Sciences. Paper prepared for: Handbook of Communications, published by Rand McNally.

SHERIDAN, T.B. (1971) "Citizen feedback: new technology for social choice." Technology Review, M.I.T. (January).

———(1971) Use of Electronic Voting Devices and Formal Models in Group Discussion and Decision Making. Unpublished draft report, August.

———(n.d.) Telecommunications System for Group Feedback: Seven Problems of Design. Unpublished progress report on ongoing research in technologically aiding the group-meeting process.

STETTEN, K.J. (1971) Interactive Television Software for Cable Television Application. MTP-354, The MITRE Corporation, June.

———(1971) The Reston, Virginia, Test of the MITRE Corporation's Interactive Television System. MTP-352, The MITRE Corporation, May.

STEVENS, C.H. (1970) "Science, government and citizen feedback." Operations Research, Vol. 18, No. 4, July-August.

———(1971) "Citizen feedback: the need and the response." Technology Review, M.I.T., January.

TAVISS, I. (1970) Ed. The Computer Impact. Englewood Cliffs, New Jersey: Prentice-Hall.

Related Policy Publications Available through the Center for Policy Research

1. Genetic Fix, by Amitai Etzioni, (Macmillan)
2. Enforcing Air Pollution Controls, by Esther R. Schachter, (Praeger)
3. Hospital Efficiency and Public Policy, by Harry I. Greenfield, (Praeger)
4. Technological Shortcuts to Social Change, by Amitai Etzioni and Richard Remp, (Russell Sage Foundation)
5. The Illusion of Equality: The Effect of Education on Opportunity, Inequality and Conflict, by Murray Milner, (Jossey-Bass)
6. Interorganizational Outputs: Case Studies of Health Care in New York City, by Robert Alford, (Center for Policy Research)
7. More Equality, by Herbert J. Gans, (Pantheon Books, Random House)
8. Engineers' Employment and Public Policy, by Trevor Bain, (Center for Policy Research)
9. Coordinating Health Cares: Explorations in Interorganizational Relations, by Edward W. Lehman, (Sage Publications)
10. Community Crime Control: An Exploratory Study, by Sarajane Heidt, (Center for Policy Research)